3 STEPS BEYOND

High-Level Questioning for Physics Mastery

By

Eric J. Burtson

Table of Contents

Introduction

Chapter 1: Why Question Higher?

Chapter 2: Purpose and Goals

Chapter 3: Relevant Autobiography

Chapter 4: Making and Categorizing Questions

Chapter 5: Forces and Motion
- Constant and Changing Speed
- Newton's First Law
- Newton's Second Law
- Newton's Third Law
- Universal Law of Gravitation
- Perpendicular Force on a Moving Object, Direction and Speed
- Circular Motion and Centripetal Force
- Centripetal Acceleration
- Relativity
- Projectile Motion in Two Dimensions
- Vector Resolution
- Balanced Forces, Statics
- Inverse Square Laws, Gravity and Static Electricity
- Forces and Motion, General Questions

Chapter 6: Conservation Laws of Energy and Momentum
- Kinetic Energy, $E=\frac{1}{2}mv^2$
- Potential Energy, $E=mgh$
- Conservation of Energy
- Momentum, $p=mv$
- Conservation of Momentum
- Impulse and Momentum Change
- Conservation of Momentum and Energy, Elastic and Inelastic Collisions

Conservation of Energy in Systems, Capacitors and Springs
Conservation Laws of Energy and Momentum, General Questions

Chapter 7: Thermodynamics
Energy Transfer Via Work and Heat Flow
Heat Engines and the First Law of Thermodynamics
Thermal Energy and Sub-microscopic Motion
Natural and Forced Convection
Tendency Toward Disorder
Entropy as a Measure of System Disorder
Entropy and the Second Law of Thermodynamics
Heat Engines and Efficiency
Thermodynamics, General Questions

Chapter 8: Waves
Waves as Energy Carriers
Problems in Wave Speed, Wavelength, and Frequency
Sound as a Medium-Dependent Longitudinal Wave
The Electromagnetic Spectrum and the Speed of Light
Properties of Waves: Interference, Beats, Polarization, Doppler Effect, Diffraction, and Refraction
Waves, General Questions

Chapter 9: Electricity and Magnetism
Voltage and Current in DC Circuits Comprised of Wires, Batteries, Resistors and Capacitors
Ohm's Law
Power in DC Circuits
Electric Charges and Fields
Sources of Magnetic Fields and Interacting Magnetic Fields
Direction of Magnetic Field Produced by a Current
Changing Magnetic Fields Inducing Currents in Wires
Plasma, Form and Function
Electric and Magnetic Fields and Lines of Force
Force on a Charge in a Field
Electric Field from One Charge
Electric Fields from Multiple Charges
Magnetic Force on a Moving Charge

Conservation of Energy Featuring Electric or Gravitational Potential Energy
Electricity and Magnetism, General Questions

Chapter 10: Nuclear Physics
The Nucleus
Photoelectric Effect, $hf = \phi + E$
Half-Life and Decay Curves
Radioactive Decay and Decay Constant
Unified Mass Unit, Mass Defect and Nuclear Binding Energy
Fission and Fusion
Nuclear Physics, General Questions

Chapter 11: Physics, General Questions

Bibliography

Cover design by C. Michael Pedersen
Art by Jonathan W. Burtson
Edited by Robert Farrell

Some names and identifying details have been changed to protect the privacy of individuals.

All questions have been written by the author. Any similarity to existing published questions is coincidental.

ISBN: 978-0-9909287-6-8

To Donna Somerville

Introduction

It was early 2005. I was standing in front of Gary Nelson's freshman physics class at Sage High School in San Diego. Gary was near me, operating an electric oscillator, generating waves on a string. He and I were co-teaching, asking students questions about the waves Nelson was making.

We pre-planned the questions to build from low to high levels in Bloom's Taxonomy. We started by asking basic factual questions about waves. Then we checked students' comprehension with more questions.

By the time we started asking *Application* questions, there was a stirring in the back of the room. Some of the students who had been the least engaged and the most off-track all year long took notice. With their purple hair and mouths full of gum, they stood up to quietly take seats in the front of the room.

As we moved on to ask how they could make similar waves in different contexts, not only were these students answering the questions, they were asking their own.

This scene repeated itself throughout the rest of the winter and spring in science, math, and English classes. But the conditions that brought about the opportunity to co-teach and to orchestrate a powerful, engaging lesson were wiped out after one year.

I often wonder what would have happened had we kept at it. The method seemed to have caught that "dark-matter cloud of lost gifted children" (Burtson, "From Theory to Classroom to Fun"). But then we dropped them.

There are those students who will fight like Roman soldiers to get to San Diego State, UCLA, or Princeton. But there are others who need a little more convincing. They have huge potential, but unless someone builds a bridge in their direction, they won't get across.

Without the structure, these students will settle, being happy enough to slough off to the back of the room, so that after compounding three more long years of negative returns, they can claim their high school diplomas, satisfy aunt Jane, then go about the business of earning a wage. A waste.

It's a waste for these kids, but it's also a waste for society. We like the stories of people pulling themselves up by their own bootstraps, but these students are different. Maybe they are more idealistic, creative, maybe motivated by a bigger picture than a degree and a professional salary.

But we live in a society where the best jobs—and the most influential—require a college diploma. In other lands, they sort students along two paths: college bound vs.

career bound. Ours needs a third: *Creatives* who need to be activated. Who knows what can happen if we get this third wire electrified?

So this book has multiple goals, but one purpose. The purpose is to get teachers in the habit of asking the kinds of questions that trigger the imagination and drive learning deeper. We call them *higher-level questions*, *rigorous questions*, and *levels of Bloom's taxonomy*. But they transcend any academic filing system. After all, the ultimate high level question is, "Why?!"

Why should this book seek to get teachers to ask these questions? Because we don't. At least, we don't do it nearly enough. I struggle with this problem daily in my own teaching. The *goals* I have for asking high-level questions are worth many times more than the effort it takes to pause and think before planning a lesson. They are. . .

- To get students interested in what I teach them,
- To provide a way for students to make sense of everything I teach them,
- To make students more proficient with the details of the curriculum because they see the big picture,
- To improve test scores,
- To have students make better projects,
- To have students feel accomplished,
- To help students understand that they are *participants* in the accumulation of knowledge,
- To plug *Creatives* into college, and into our economy,
- To establish lifelong habits of thinking and being creative,
- To influence our culture to value thinking and creativity, and those who think and create,
- To influence our culture to be more productive,
- To influence our culture to plan for the present day and for a compelling future,
- To have fun.

Higher-level questions, whether *Analyzing*, *Evaluating*, or *Synthesizing* make people make connections. The value of this should be obvious. But making such questions is in itself *Synthesizing*, the highest-level task. It is even harder when you try to make such questions address narrowly defined points. So, while I might have filled a swimming pool full of general questions, I decided it was worth the effort to first establish a set of detailed standards, then make one *Analyzing*, one *Evaluating*, and one *Synthesizing* question for each.

In so doing, teachers could have rigorous questions for every standard they teach. And, with multiple questions per standard, they could have access to high-level questions on a daily basis.

But I don't want this book to be just a data bank of questions. It took a lot of practice for me to get good at writing rigorous questions. Using such questions is itself practice. So, in addition to the goals I listed above for asking high-level questions, another goal is that teachers learn to make rigorous questions at will. This will help everything else happen.

Finally, the summative high-level questions are indeed the most powerful. Therefore, at the end of each major section of standard topics, I include my favorite high-level summative questions. This first book focuses on physics. Have fun.

Chapter One: Why Question Higher?

SOME HISTORY

What if we only questioned low? What if all we asked students to do was memorize definitions of terms? This base category in Bloom's is called "knowledge". What's wrong with that?

In *Surely You're Joking Mr. Feynman*, the author talks about just that situation. It's crazy truth that sounds like fiction, involving physics students and their education at a university in the developing world. Their tests only asked them to repeat definitions of physics terms they heard in lecture. Feynman saw the futility in the system immediately. The students didn't "know" anything. They were just parrots.

Look up "feynman-brazil-education". You will see what happens when people stay pegged to the bottom level of Bloom's Taxonomy. Better yet, read the book. It's hilarious.

I'm not saying there's no place for *definition*. The opposite end of the spectrum would be just as bad. What if everyone's schooling involved only the act of *creation*, but without the ingredients to make anything—bricks without straw, Einstein without light? We'd be playing Dungeons and Dragons.

Before we start feeling superior, we need to look at our own situation. In 1983 The *National Commission on Excellence in Education* wrote *A Nation at Risk*. The report emphasized the urgently lagging state of American education, reporting "an incoherent, outdated patchwork quilt," of teaching and learning in our country. "We have," it said, "in effect, been committing an act of unthinking, unilateral educational disarmament."

This got people's attention, but it didn't necessarily change things, as national standings showed. But people did react. In *The Teaching Gap*, Stigler and Hiebert reported on the results of a 1995 video study of teachers across the globe, *The Third International Mathematics and Science Study* (TIMSS). The authors show that both Germany and Japan's math instructors tend to ask questions that encourage the development of concepts 80% of the time, versus questioning for skill practice 20% of the time. In America, that 80/20 ratio is inverted, which, they say, results in the international performance gap. *They spend 80% on concept development to our 20%*!

If you don't already know, and as I'm sure you suspect, "skills practice" resides in the bottom half of the taxonomy, while "concept development" dwells in the upper half. Maybe we in America haven't isolated ourselves to *just* the first floor, but the view only gets good on floors 4, 5, and 6—and we are a long way down from there.

About the same time *The Teaching Gap* came out, another reaction overtook it: the standards movement. As a nation, we turned toward the belief that we could get ourselves on the path to progress by identifying critical points of content, teaching these, and testing the results. The results would inform our "instruction" and get us caught in an improvement spiral upward.

What more convenient and manageable way to compare results than to make multiple choice tests? And what other instrument virtually guarantees structuring everything around *recall*, *description*, and *problem solving* at the expense of *analysis*, *critique* and *creation*? So, in an effort to close the gap, *we locked it in*.

That's pretty much where we are right now. The Next Generation Science Standards might be an improvement. I especially like their "cross-cutting" standards, but we need more.

TEXTING

I teach International Baccalaureate (IB) Physics. IB has lot of standards, yet it also has a big lab requirement. No matter what your curriculum, labs are high level, especially when they are student-designed. So I am happy with IB.

Of course, every class needs a textbook. The district I work for adopted the same book for Advanced Placement and for IB Physics. It's not a bad textbook, but I wanted to quantify how it was doing. Therefore, I threw dice and randomly chose three chapters to analyze. Each chapter had about a hundred questions for student practice. I went through each and categorized the questions from 1-6 through Bloom's taxonomy, with six being the highest. Here are the results for the 320 questions I categorized.

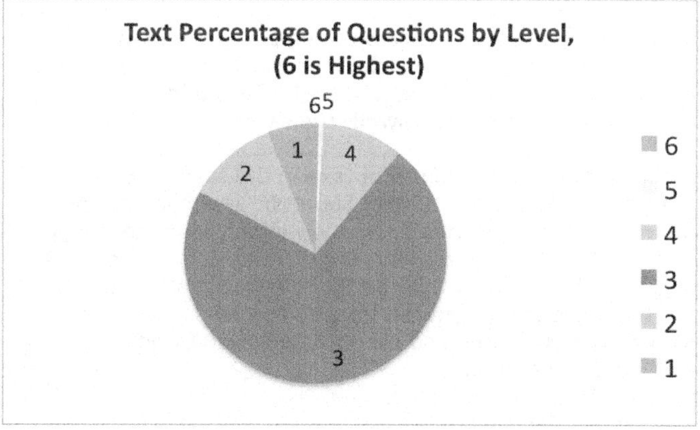

Those indiscernible slivers are levels five and six. *Level five* captured two questions, *level six* got one—*out of three-hundred, twenty*. Almost three-fourths of the questions were at level three, knocking on the door of higher level, but not quite there.

These are the questions most commonly recognized as physics questions. They seek one right answer, usually through the application of some learned formula. You might be asked to jump through hoops, use one equation to get part of the answer, and plug that part into another equation. It might have a long set up, "a train is moving south at the rate of 100 km/hour. A second train is going north at the rate of. . ." They're not bad questions. Questions at each of the six levels have a place in pedagogical use, especially level three for future engineers.

But if part of the purpose of instruction is to make the student make connections, see differences between concepts, and clarify each concept in his mind; if part of the purpose is to sharpen the use of common sense and informed opinion; if part of the purpose is to have the student imagine, and draw upon all resources to make a new thing, then an out-sized reliance on level three is not acceptable.

THE SAGE EXPERIMENT

Going to higher levels worked at Sage High School. As I wrote in the *Introduction*, our Principal, Dolly Samson, brought together several elements to pull her school forward. She acquired a full slate of content administrators, she redesigned the entire ninth grade into a small community emphasizing *rigor*, *relevance*, and *relationships*, and she had us work with teachers to fashion questions at all six levels. To drive the point home, she gave the year a theme: *Think Like a Genius*. After all, 2004-05 was the 100th anniversary of Einstein's big three papers. At Sage High, higher-level thinking was encouraged.

Like any experiment with a lot of variables, it is hard to determine the most important factors. Some of it, of course, was just plain teacher buy-in. Most of those who worked with the freshmen class voluntarily moved to the 100 building to do so. Others contributed passionately and creatively towards the cause.

George Rochambeau for example, taught freshman physics in the small FOCUS community (Freshmen Oriented for Creating Ultimate Success). He conceived of a school-wide professional development featuring a series of skits in which teachers acted out three of Einstein's 2005 papers: *relativity*, *the photoelectric effect*, and *Brownian motion*. Throughout, the actors elucidated the science via a series of low and high level questions. The audience was all the school's teachers, seeing their colleagues explain difficult concepts clearly, and becoming engaged by good questions. It was on point, fun, and brilliant.

Our work as science administrators took place in the prep rooms and in the classrooms. We used a coaching cycle in which we planned a lesson together; infusing it with questions that went up through the higher levels as the lecture progressed. The content administrators delivered the first lesson, modeling the questioning live and from script (those high level questions that we planned as *must haves* for the demonstrations). Later, we debriefed the lesson with the teachers, and co-wrote a second lesson for higher-level questioning. This next lesson we taught together—two teachers in one classroom! We debriefed again, co-wrote a third lesson, and gave the lecture completely back to the teacher. We observed, gave feedback, and debriefed again, parting ways until the next coaching cycle.

I feel these interactive questions were the most effective at activating student thinking. Not only did students consider the questions themselves, but they witnessed their peers answer in ways they weren't used to.

We also encouraged teachers to pepper their tests with some higher-level questions, maybe have them lop off ten redundant *recall* and *problem-solving* questions for one *Analysis*, one *Evaluation*, and one *Synthesis* question.

All the work paid off. Sage High jumped 42 points on California's Academic Performance Index (API). This was the highest move up among the 17 high schools in the district.

I am convinced a part of that gain was due to the emphasis on higher levels. Asking students to think in six ways rather than three simple ways helps even for tests that gauge progress in the three simple ways. In my years playing bass, I have noticed that the best musicians in bass, guitar, and even drums often are those who have the strong musical background of learning piano. It's a similar phenomenon.

TSK TSK

Getting back to the book—the textbook I analyzed above—that was a little unfair of me. Physics is a lab science. Doing lab work can be the highest level, *Synthesis*, especially when students design their own investigations. Maybe the text authors knew this and were happy to delegate the highest levels to the lab.

Still, one *Synthesis* and two *Evaluation* questions out of 320 seems more than a little stingy. Imagine if there weren't even labs! About 15 years ago, when I attended IB Physics training, I heard there was a movement among AP Chemistry teachers in the southeastern US. They felt so overwhelmed by the content requirements that they started *not* doing labs at all.

I bring this up, because, when the pressure is on, we humans like to revert to things we can control, like curriculum pacing. Ejecting labs is a good way to get on track, but at what price? This is one example of how easy it is to regress away from the higher levels.

MARS IS HIGH

One person who does not shy away from the higher levels is Elon Musk. In November, 2015 one of my students' parents arranged a tour of Musk's SpaceX facility in Hawthorne, California—just me, some parents, and three students.

One of Musk's right-hand men took us through the facility. It was indescribably intriguing, seeing people take over the task I saw NASA do as a kid. The guide talked about a "million parts," and I saw all kinds of pipes and manifolds spread across office desks like desultory detritus, but with a purpose. These guys took over the NASA vision and made it bigger, in a casual determination that nevertheless carried unstoppable inertia.

Our guide told us fascinating details, like how the thrusters exceed the power output of a megalopolis. But he told us "Elon's always talking about Mars."

Elon Musk and Stephen Hawking are two people who share the vision, the understanding, that if humans are to go on, we must get off the planet. Why? Diversification of habitat. One day another giant asteroid will hit the earth. Last time that happened 99% of the species died with it, along with every single dinosaur. Lucky for us, because we evolved as a result. Next time, we are likely to be in the unlucky 99%—unless we are somewhere else, or five, ten or a hundred different places. Mars is the logical *next* place.

We keep listening to deep dark space, and we don't hear anything intelligent out there. Is it possible we are the only ones? We must go on. Evolution has baked into us a powerful need to procreate. It did this by walking our genes through an incredible gauntlet of challenges, locking in a hardy tenacity for survival. But we didn't have to think our way through to our current design. Nature and chance brought it about. Perhaps that's what makes Elon Musk the ultimate male: He is not settling to be one random link in the chain; he is actively looking for a way for all our genes to survive—forever.

I realize I am getting a little grandiose here but I believe we can help it happen if as a species—we activate our youth's *Knowledge*, and their *Comprehension*, and their *Application* and their *Analysis* and their *Evaluation* and their *Synthesis* to be real, connected, coherent, deep, and unbounded on the upside, because *up* is the only way we go from here.

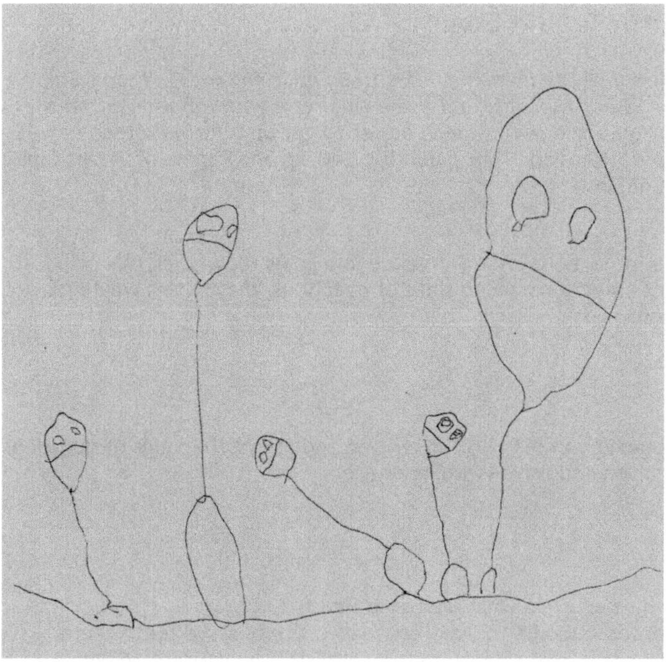

Chapter Two: Purpose and Goals

> *I know the tendency of the human mind is to do anything rather than think. But mental labor is not thought, and those who have with labor acquired the habit of application often find it much easier to get up a formula than to master a principle.* (Harman, "The Scientific Letters and Papers of James Clerk Maxwell Vol. I," 662-674)

In case you skipped the intro, here are my purpose and goals for writing this book. If all teachers everywhere asked a balanced slate of questions, maybe this would move us closer to reaching these goals.

PURPOSE

The purpose is to get teachers in the habit of writing and asking the kinds of questions that trigger the imagination and drive learning deeper.

GOALS

- To get students interested in what we teach them,
- To provide a way for students to make sense of everything we teach them,
- To make students more proficient with the details of the curriculum because they see the big picture,
- To improve test scores,
- To have students make better projects,
- To have students feel accomplished,
- To help students understand that they are *participants* in the accumulation of knowledge,
- To plug *Creatives* into college, and into our economy,
- To establish lifelong habits of thinking and being creative,
- To influence our culture to value thinking and creativity, and those who think and create,
- To influence our culture to be more productive,
- To influence our culture to plan for the present day and for a compelling future,
- To make school fun for teachers and students.

Chapter Three: Relevant Autobiography

Chapter Four lays out exactly how I write higher-level questions. Chapter Three gives you my professional life journey, including my insights into where I encountered and where I invoked higher-level questions. I include results, however anecdotal. If you are not a fan of autobiographies, you might want to skip this one. Either way, let's meet up again at Chapter Four. I start with a brief sketch of my professional journey, followed by a string of realizations about the value of high-level work.

THE ROUTE

I stepped out of college into an engineering position at a small biotech. Five years later I launched into science teaching, starting at middle school, then bouncing between high schools.

I soon got my dream job, teaching IB Physics in San Diego, "Sakura High". I loved the curriculum, the students, my teacher buddies. I had some success. After a few years I got the leadership bug. I wanted to make a change, turn schools around, move towards excellence, all that.

Thus began a ten-year odyssey of wins and losses, triumphs and failures. Over that span I was at five different schools in two different districts, every year was a new position, back and forth between being a teacher and a low-level pot-smoker-chasing administrator. I was a science VP at one school, a *WASC* coordinator somewhere else, a new kid on the block, a vice principal.

I learned a *lot* about leadership, good and bad, what it takes to make change happen, and how easy it is to let things roll on in a state of semi-dysfunction. I have seen tremendously successful reform, and I have seen the same reform abandoned and let go feral. I have served under twelve principals, some amazing, some still developing.

Eventually, I left education entirely. After a couple of years doing something else, my friend Tabitha Breen gave me a call, saying they really needed a physics teacher back at Sakura High, so I decided to get back in.

To my surprise, I was energized. The effect of all that experience was better connection with my kids. Usually, I'm focused on the subject I teach. Now, I see the people too. I also had the simple realization that my greatest influence can be on my classroom and those who pass through it—regardless of the craziness outside those walls. In fact, any change has to start inside those walls—hence, my new excitement about rigorous questioning. However, I still don't decorate well.

EARLY RIGOR

When I was in high school, a few of my non-academic friends called me "By-the-book-Burtson." It took me a while, but I eventually learned that the fun stuff, and the higher stuff, was not in the book.

High school didn't give me any real sampling of rigor. My science and math classes were standard, hitting a ceiling at level three, *application*, problem solving. English essays probably poked into the higher levels, but that's about it. History was a joke: multiple choice factoid tests—the only class I ever cheated in. My good friend Kyle Crimson just waited until the football coach left the class to do whatever he did. Crimson strutted up to the file cabinet in the front, lifted the test, and shared it.

College was almost the same deal, not the cheating, but the mundane. I got my first ever *F* grades in a few of my engineering courses. But I also excelled in several others. It seems that I did my best work when the assignments were higher level. There weren't enough of these, however.

Here are four exceptions that made the deepest impression on my memory. I don't remember much else from college work. I suspect I remember these because they were more challenging.

The way the professors structured these assignments worked, in some way, to teach me deeply. If they were not all-higher level, they were very close. But there is value in every level, so I describe each of these four impacts.

I have a soft spot for level three, *Application*. Like the virtuous pagans in Dante's *Inferno*, *Application* almost made it in. Future engineers need to spend a lot of time there. It's how they learn to persevere and solve problems. After all, it is the *highest* of the low level questions.

Story one of four from college is all about that. I remember that my Statics class assigned a huge problem set every week. I would bury myself in the basement of the engineering library, taking water breaks and finishing the weekly load after 10-14 hours of straight work. It was all level three. But after being beat up by it, I started to understand that torque happened at every point along a crane's length.

Story two was a random assignment in an upper division programming class. The professor gave each of us two handouts, one that described finite element analysis on an airplane wing and another that outlined the project. We were supposed to figure out how to program to solve a stress problem with the wings. This project was very difficult, but right on the cusp of higher level. There was a best solution (level three), but there was a lot of comparison and breaking down of information (level four).

I remember how opaque the problem was. I sat in the quietest place on campus, the biomed library. I looked back and forth between the two documents, trying to find connections and clues and sense. It took hours. Eventually a hint of understanding came, then another, then the compounding of insight. I left that place knowing exactly how to do it. To my surprise, my peers hadn't cracked it, and I was a kind of hero among them.

Story three was a project in a soils class. Have you ever wondered how the piers that go from shore to sea stay in tact after decades of getting smacked by waves? My project was to investigate how they pounded the poles into the ocean floor so that they could continue to support the pier regardless of the weather and the water. Again, it was complicated, and again, it wasn't clearly one level. To understand it, you had to follow the physics, the application; but the number of strokes of the pile driver, and the masses—it was more like a recipe than a formula. I remember thinking, really? So, it had a flavor of *analysis* and *evaluation* too. It drew me in so that again, I felt like I owned it at a conceptual level. I delivered the presentation with authority.

Story four was pure level 6, *Synthesis*. Our thermodynamics prof gave us a major assignment comprised of two big problems. I was deeply involved in clubs. I kept too busy. Shamefully, I procrastinated this assignment to the point that I had to salvage it by choosing one of the two problems to complete by the next day, when they were due.

I thought I would at least finish the easier problem, designing a radiator that would cool to meet certain specifications. I attacked it, considering fin size and density and several other variables. I optimized it. It worked well. Whew. One problem done, not a major loss.

As it turned out, that problem was the big one, so much so that it really didn't matter that I didn't do the other one. Again, my peers were impressed. I thought I was just stupid for procrastinating.

I share these stories because they made the biggest impression on me, and, because they were the closest things to higher-level work I did in college. I have to say each of these four experiences were also empowering. I remember leaving the classroom after my piers presentation stuffing the posters and models into the first trashcan I saw. My partner was shocked that I would do that. I did it because it was all internalized. Done, and a part of me forever.

ACTIVE RIGOR

I reflected on my pre-teaching professional life, to see how much of my engineering career called on higher-level thinking. It seems obvious that it would. But not necessarily.

For example, my buddy Ross got an internship in college. He worked a full summer for a major industrial firm. One of the engineers he worked with was near retirement. This guy got his position way back during the space race, when we needed an influx of engineers—but maybe he shouldn't have been one of them!

The guy asked Ross for some help on a project that kept him baffled for quite some time. All this person was supposed to do was read numbers from a printout and reformat the numbers so they were no longer in fraction form. The numerator and denominator each had polynomials in them.

The problem was arguably not even level three because, it could be argued, his "computation" was more like a simple restatement (level two). You would think the guy would have had some mentors in that big organization.

Conversely, I worked for a very small company. I was the only engineer. We were making simple medical devices, modifying existing models that sold well, but also creating brand new products all the time. Most projects we abandoned when we realized they would interest only niche markets.

I kind of expected that someday, when I became an engineer, there would be some guidance, a book or something! I imagined that there would be standard functions outlined in a procedural manual, detailing work I would be qualified to do because of my degree. Maybe I thought that way because that's how college prepared me—ready to roll up my sleeves and problem solve, level three style!

Sure, there was some of that, but the important work needed something higher. There was one new project that presented problems. The skilled technicians who made the devices tweaked the controls on the ovens and the motors. It was all trial and error. Some days they got things right, and the development progressed. Other days didn't go well. They needed a way to make the process more predictable.

So, I was thinking about it, pencil on paper. I eventually found a solution—seven equations and seven unknowns. It worked. Later they hired more engineers. My solution became something they could easily apply again and again. It was a level three *Problem-solving* application—but not that first time I found it! That was *Synthesis*.

RIGOR, INTENTIONAL USE

The first time I started intentionally applying the taxonomy I wasn't aware that I was. Helen Simmons, teaching 8th grade science one door down the hall from me, got me going with project-based learning (PBL). PBL worked for my students and me—they started making things in class, and learning too! And it was way more fun than all the by-the-book stuff I thought I needed to do 100% of the time.

Of course, PBL touches the upper levels, and can go there aggressively, with the right project description. All I knew at the time was that PBL worked. In fact, some of the questions in the later chapters are structured as mini-labs or mini-projects.

Science teachers assign high-level work automatically when they do labs. Sure, cook-book style labs are only at the *Application* level, but student-designed experiments take it all the way to *Synthesis.*

Maybe that's why, when I was a science administrator, our trainers were so crazy about *inquiry*-based science instruction. See Douglass Llewellyn's book *Inquire Within, Implementing Inquiry-based Science Standards* (Llewellyn, 2001)

After leaving education and later coming back to the classroom, I wasn't intentionally writing high level questions for my students. I was already doing labs and competitive projects. The IB curriculum already felt rigorous enough. On top of that, I re-introduced writing at the beginning of each period.

I liked to avoid the mathematical problems in these prompts. I wanted words, and I wanted them to wrestle with concepts. I had already acquired a facility with writing higher-level questions, and that process probably became automatic with these writing prompts.

Upon writing this book, I reexamined those prompts to test my hunch that I had been making rigorous questions all along. As it turns out, most of them were.

I got back into the habit of intentionally composing high questions just this year, when Percival Green started teaching AP World History. Our principal wanted every student to take the class. There would be no separation of students between this course and regular world history. Percival wanted everyone to succeed, so he formed a small committee to consider the factors that would support this goal.

One afternoon he produced a list of about twenty possible causes for last year's low scorers. Jenna Baron, who also teaches AP World, got the list too. When Percy asked us both what we thought, we had already circled the one word, *rigor*.

Later, I dusted off my old professional development tools from my Sage High days, and we did a high-level question writing session with the AP World teachers. By then I realized that I needed to become intentional about writing these questions for my own students.

I started putting one *Analysis*, one *Evaluation*, and one *Synthesis* question at the end of each test.

Here is an *Evaluation* question I put at the end of my regular physics (non-IB) test:

> Cables attached to a motor move elevators up and down inside a building. If the cable breaks, people usually die. However, some people say that if you could jump just before the elevator hits the bottom, you might survive. Do you agree with this idea? Why or why not?

I include this one because it was an example of a question that caught a lot of misconceptions. Here was one answer, "Yes, because if you jump right before the elevator hits, you won't suffer from as much impact, meaning you won't hit the ground as hard."

Hmm. Well, we hadn't had our unit on *energy* yet, but still, they could do speed calculations fine. They could also demonstrate an understanding of acceleration and speed caused by gravity. *Without* that question I would have thought everything was hunky-dory. By *giving* the question I caught a misconception I could later address. And, it *did* cause some students to think:

Some students also confuse weight with impact force. This student looked like he was going to make that mistake, but instead, he came up with a more logical answer than the previous student. He did a speed calculation based on a reasonable starting height he selected. As a result, he could see that a last second leap would not have much effect on the falling person's final speed. Plus, by his phrase "net gain of forces" I could see he was developing concepts that are directly related to energy (*Work* = *force* x *distance*).

When I asked my IB Physics students what they thought about the questions I put on their test, they had indeed noticed the difference. They said the questions were "interesting". They said they liked them, and the questions caused them to think. I later heard back from my friend Tabitha Breen that some of the students said I was "trying new things".

Here's my last story. One of my most capable first year IB students is having health problems. In preparation for a possible long-term upcoming absence, she, her counselor, parents and teachers had a meeting. We all agreed on ways to help her. As we were wrapping up, she asked for one more thing: that the first few minutes of my class could be recorded daily so she could hear the dialogue. She said that was her favorite part of class, and she didn't want to miss it. Yay for higher-level questions! Now the question is, *Who's going to record it?*

Chapter Four: Making and Categorizing Questions

When I cut my teeth on rigorous questioning, my peers and I went back and forth between two resources, Bloom's original list of categories and words, and Anderson et al's revised list. These links will give you similar resources: Bloom's original: http://www.krummefamily.org/guides/bloom.html
Bloom's revised: http://thesecondprinciple.com/wp-content/uploads/2016/01/Anderson-and-Krathwohl.pdf

If you want to dive deeply into the two books, you will get better context and detail. Some people prefer Bloom's original, and some prefer Anderson, Krathwohl, et al: http://www.amazon.com/Taxonomy-Educational-Objectives-Book-Cognitive/dp/0582280109/ref=sr_1_3?s=books&ie=UTF8&qid=1453854894&sr=1-3&keywords=Benjamin+Bloom and http://www.amazon.com/Taxonomy-Learning-Teaching-Assessing-Educational/dp/0321084055/ref=sr_1_4?s=books&ie=UTF8&qid=1453854785&sr=1-4&keywords=bloom%27s+taxonomy

Whether you get the books or follow the links above, you will encounter a bunch of words that help describe each level. The names of the levels themselves describe the goals for the questions at each level. Anderson modified the levels by changing the descriptors from nouns to verbs. She and her colleagues also switched Bloom's two highest levels. Under her scheme, *Create* comes above *Evaluate*. After all, shouldn't the playwright be more important than the critic?

I like that, but I took it a step further. I made each category a verb representing what you want to *see* a student do. For example, her hierarchy from top down goes from *Create* to *Evaluate* to *Analyze* to *Apply*. That's good, but at the lowest two levels you have *Understand* and *Remember*. Those are not words that imply action that you as a teacher can see your students doing.

My re-titling of these terms for this books goes like this, from top down: *Making, Justifying, Analyzing, Solving, Explaining*, and *Recalling*. For example, Bloom would call for *Comprehension* at level 2. Anderson would tag that as *Understand*. I prefer *Explaining*, because I can ask a student to do that. Here are all the terms for comparison, with higher levels bolded.

Level	Names of Levels for Questioning, Highest on Top *(Note the re-ranking of the two highest levels)*		
1-6, *Six is high*	*Bloom*	*Anderson, et al*	*Three Steps*
6	**Evaluation**	**Create**	**Making**
5	**Synthesis**	**Evaluate**	**Justifying**
4	**Analysis**	**Analyze**	**Analyzing**
3	Application	Apply	Solving
2	Comprehension	Understand	Explaining
1	Knowledge	Remember	Recalling

The original and revised versions give more words to clarify each category. For example, *Analysis* includes *breaking information into parts, comparing and contrasting, differentiating*, etc. Anderson, Krathwohl, et al even expand the taxonomy into a two dimensional grid, subdividing the categories. The describing words fit nicely into that grid.

I prefer to simplify the question-writing process by holding on to a few differentiators between levels. If you want more guidance than that, I recommend printing out Bloom's and/or Anderson's lists.

Below, I pose some possible physics questions, then reason through the categorization. This can provide you some practice, as you anticipate the level I would place a question. After that, I sum my process for each level, which will help you start to make your own questions. The rest of the book has the questions I made for a detailed set of physics standards as well as questions that are broader than the specific standards.

JUMPING INTO CATEGORIZING

What level are these questions?

> Q1: If you were in a car accident, would you rather it be a totally elastic or totally inelastic collision? Why?

Recalling? No. While you might have to remember what *elastic* means, the task requires some thinking.

Explaining? You will definitely be explaining your answer, either verbally or in writing, but it is not likely that *car collision impact on the driver* was specifically gone over as if it were a general topic. Rather it is something that requires some application of concepts. So it is not *Explaining*.

Solving? Application is not enough to anchor this down to the s*olving* level. There is no one right answer. The possibilities make this question richer and higher. The respondent must apply various concepts to various scenarios and make comparisons.

Analyzing? We're getting close. The respondent is making comparisons and breaking things down as is characteristic of the *analyzing* stage, but his ultimate answer may be more of an opinion. A cut-and-dried conclusion may not be reachable. For example, he may think that a little bit of crumpling and energy absorption would be beneficial as long as there is some smoothness of impact that a partially elastic collision could bring. Others might recognize valid reasons for a crash being more purely elastic or inelastic, especially when considering the masses of cars they might be driving compared to the masses of cars they might hit.

Justifying? Bingo. The question requires justification for all the reasons I mentioned above. And, while the respondent is *making* a reply, it is likely not a *model* for assessing preferential scenarios from a set of possible accidents. If the respondent produced that, kudos! But it was likely not the intention of the question writer.

Making? No.

> Q2: Think about the ways we make our houses able to conserve as much heat as possible on a cold winter day. In writing, compare these methods with the statement, "most processes decrease a system's order, and eventually energy is spread evenly." Annotated sketches are welcome.

Recalling? No. Clearly there is too much going on with this question.

Explaining? Nope. Requires deeper thinking.

Solving? Closer. The respondent will be applying concepts, but there is no one right answer. Also, he'll be making a lot of comparisons.

Analyzing? Yes. There'll be some breaking into parts and examining relationships.

Justifying? This could get to *justifying* if actual plans or a set of proposals were offered and the respondent was asked to choose one and defend his choice.

Making? It could easily get to *making* too, if the question asked the respondent to formulate a plan.

> Q3: Design a situation in which the Doppler effect results in a smooth change in sound frequency.

Recalling? No, go straight to *making*. The word *design* pretty much identifies the question as *making*.

Explaining? No. →
Solving? No. →
Analyzing? No. →
Justifying? No. →

Making? Yes!

> Q4: On paper, sketch several regions of magnetic fields so that if a charge entered from one side, its path would trace an interesting pattern or shape. Be ready to explain how you did it.

Recalling? No, not simply recalling something.
Explaining? No. This is a new situation.
Solving? No. There is not one right answer. This would be the level if the problem called for making a particular shape.
Analyzing? No, the problem did not present itself as something wherein a lot of comparisons needed to be made. It's more applying what you know to *make* what you want.
Justifying? No opinions. It's *his* plan.
Making? Yes, the respondent is *making* something.

> Q5: What is the right hand rule for a moving charge in a magnetic field and how do you apply it?

Recalling? It could be this level, but there is an element of explaining.
Explaining? Yes, asking how it is applied makes this question an explanation.
Solving? If the questioner gave a scenario that anticipated one right answer using the right hand rule, then yes.
Analyzing? No.
Justifying? No.
Making? No.

> Q6: You shoot a 5.0 kg cannonball through a 2.5 meter barrel, angled up at 50° from the horizontal. It takes 0.070s to get through the barrel. Find the following:
> a) The cannonball's exit speed,
> b) The average force on the ball in the barrel,
> c) The ball's highest point in its path (neglecting air resistance), and
> d) Its range.

Recalling? No.
Explaining? No.
Solving? Yes. Even though there are several answers, they are answers to specific sub-questions, and there is only one possible answer for each.
Analyzing? No. Though there may be a lot of mental gymnastics going on in the solution, like making comparisons between this problem and others requiring the same algorithms, this is not *analyzing*. *Solving* uses learned information in new situations, like this one.
Justifying? No.
Making? No.

Q7: What are the approximate wavelengths for red, green, and blue light?

Recalling? Yes! I had my physics students memorize a few of these the other day. Then I verbally quizzed them on these factoids. Though these values could easily be looked up, they are handy to know because they are relevant to a lot of other work.
Explaining? No.
Solving? No.
Analyzing? No.
Justifying? No.
Making? No.

This last example is a question transformed from low level to high level. This is a revision of Q7.

> Q8: Cassandra says that she wishes she could see infrared colors as well as all those in the visible spectrum. Darsha says she would rather see ultraviolet. Who is right and why?

Recalling? No.
Explaining? No.
Solving? No. There is no one right or best answer for this question.
Analyzing? No, but there will be analyzing done upon answering this question.
Justifying? Yes. The question requires comparisons and judgments. After all the evidence is laid out, a final answer will be based on opinions. Obviously, there are good reasons to see in each of the two proposed sub-spectrums of color. The question is rigorous because it causes the respondent(s) to think a lot from divergent sources. This could form the basis of a class debate at the end of a unit on light. Some thoughts this question might invoke include the following: *Ultra violet would give the seer more resolution, like a blue-ray dvd. . .Or would it? Maybe the extra function of the eye would affect its form, sacrificing space for rods and cones to accommodate the UV detection tissues. . . Infrared vision might allow people lost in the wilderness to locate sources of heat. It might also allow them to see dangerous leaks in a factory. . .*Likewise, *UV might allow seers to tell when conditions are wrong for sunbathing.* Or, debaters might research the breadth of IR vs. UV frequency ranges, determining which power would be a "better value". This could go on and on, building and strengthening understanding of light.
Making? No.

JUMPING IN TO MAKING QUESTIONS

Ok, if those examples weren't enough to get you started writing your own high level questions, here is my process, summarized. I'm not claiming it to be as comprehensive as Bloom's or the revised Bloom's method, but it works for me for writing physics questions.

When you've made a few, email me with your questions and how you categorized them and I'll match it to my ranking. Just kidding.

I offer no advice on writing questions at the lower three levels. We learned that by example, k-16+. The main differentiator to keep in mind is *divergence* vs. *focus*. This alone can guide you higher. The lower levels want black and white answers. The highest of the lower, *solving*, is going to make you work for it, but it's still lower level, and seeks one right answer.

Here is the higher level, starting one step beyond *Solving*:

Analyzing, with its comparisons and contrasts, breaks things apart and examines them. It seeks to recognize the essence of the differences, in form and function.

. . .two steps beyond *Solving*:

Justifying compares and contrasts as well, but it judges value. It invokes opinion, and there can be many opinions, following tangents, making connections, seeking evidence.

. . .and three steps beyond *Solving*:

Finally, there is *Making*. *Making* creatively or divergently applies what is already known to make something new. It results in models or devices, conceptual or material.

So, if you want to make higher-level questions, you can start with a particular standard, like *Snell's Law*, or a bigger topic, like *Electricity and Magnetism*. Then think.

If you want a *Snell's Law* question, let's target *Analyzing*. What can we compare? What can we differentiate? Maybe we can start with the devices that already exist that use refraction, and ask for a comparison:

> Compare and contrast how lenses and fiber optics take advantage of the fact that combinations of two media have different refractive indices.

If you want to try *Justifying*, try to make a heated debate among students or in each person's mind.

> Eric says his glasses give him headaches. He is looking into making his entire car windshield a lens that matches his prescription. Is this a good idea?

If you want a *Making* question, have the students build something with ideas or real materials.

> How might you make corrective lenses that can be instantly adjusted to match a user's prescription?

Or

> Arrange these glass and plastic pieces to make the fattest possible spectral dispersion band using this white light source and 0.3m square for your assembly.

The same process works for the general topic *Electricity and Magnetism*. I think it is easier and probably more beneficial to make questions for broad topics. The only drawback is that if you rely exclusively on the broad topics, then you have to wait until the end of the unit to do the high-level dance.

Analyzing

> Think of lightning and think of getting shocked by touching a doorknob. How are these two events the same and how are they different?

Justifying

> Shelby says we could save a lot of money on electricity if we just captured the electricity of a thunderstorm. Her plan is to put giant capacitors in

underground insulated chambers and connect them to lightning rods that are charged slightly opposite that of the clouds, to coax the lightning to strike. As soon as the lightning hits, it would charge one giant plate of her capacitor. A mechanism would react quickly to disconnect the lightning rod so that the charge would be captured. Can this work? Why or why not?

Making

Aside from examples already discussed in class, name three practical (or fantastic) uses of a circuit comprised of a charged capacitor and a wire loop.

Well, there you go. I started with examples of categorizing and ended with examples of writing. The rest of the book is a bank of questions going one, two, and three steps beyond *Solving*. My hope is that you find these useful. . .and stimulating enough that you will write even more questions. Good luck.

Chapter 5: Forces and Motion

Constant and Changing Speed

ANALYZING

Sketch two graphs. The first depicts a constant speed *v* of 40 m/s on the vertical axis. It has time *t* on horizontal axis running from 0 s to 20 s. The second graph has a speed that builds steadily from 0 to 80 m/s over the same period of time. Explain how you can use the graphs to find the distances travelled in 20 seconds and how those processes are similar and different.

JUSTIFYING

Your family is driving from Los Angeles to San Diego. Your buddy Joe from Physics class says it is best to drive at a constant speed to get there quickest. Nora disagrees. She says her engineer dad drives differently. She says her father always looks for the most efficient way to do things, therefore his frequent accelerations are meant to minimize trip time. Who's right, Joe or Nora? Justify your answer.

MAKING

Los Angeles is 120 miles north of San Diego. San Francisco is 385 miles north of Los Angeles. Design a bullet train trip that starts at 6:00 a.m. in San Diego and ends at 8:30 a.m. in San Francisco after stopping in Los Angeles. Allow time for the train to steadily increase to a constant speed and to steadily decrease to a stop. Also allow a reasonable loading and unloading time at each city. Specify the times, speeds and waiting periods for every departure, travel, arrival, and loading time so that the train stays on schedule.

Newton's First Law

ANALYZING

A pendulum keeps swinging back and forth. The only time it is not accelerating is when it is in equilibrium, which occurs at the center position. Sketch a pendulum at three different positions: a. equilibrium, not moving, b. equilibrium, moving, and c. not at equilibrium. Explain how each position demonstrates Newton's first law of motion. You may use your sketches to help your explanation.

JUSTIFYING

Maria and Marta have different ideas about how to keep people in houses safe during earthquakes. Maria says the best solution is to drill the foundations of the house into solid rock. Marta says it is better to float a house in water. Assuming tidal waves are not a problem, whose solution is better? Why? Use Newton's first law along with your explanation. You may use sketches to help your explanation.

MAKING

Imagine you are a scientist and you are going on a trip. You have a special material that must travel inside a spherical (round) shell. The material must stay in a jar at the center of the shell. You may stabilize the jar by connecting it to the shell with pieces of string. Make a sketch showing how you would do this, but also explain this in writing, including how this task relates to Newton's first law of motion.

Newton's Second Law

ANALYZING

Can friction be a force in Newton's second law? If not, why not? If so, give me an example describing how you would calculate it.

JUSTIFYING

Josue thinks if the Olympics could be held on a planet with slightly higher gravity, we could achieve new records in sprint races. He claims that the extra weight we would get would allow us to accelerate quicker as we push on the track. Is he right, could we run faster on a bigger planet?

MAKING

Design a device to provide a constant force for a short period of time. It will be used to launch bicycles. Explain the role of Newton's second law in your design. Include a sketch of your device.

Newton's Third Law

ANALYZING

Chad just jumped out of an airplane but has not yet pulled his parachute. He has also not yet achieved terminal velocity. A seagull accidentally flies into the right side of his body. Make two sketches. The first will be Chad and the seagull at the instant they collide. It should include arrows representing the impact forces of each on the other.

The second sketch should be a free body diagram of just Chad as the seagull hits him. Explain the similarities and differences of the two sketches. Also, explain how Chad's motion will be different after the collision.

JUSTIFYING

Binu argues with Sage about the ability of Martial Arts champion Hui Gao to punch a man so hard he launches him across the room. Binu says it is possible. Sage says that even if Hui could do it, he would fly across the room too. Who is right and why?

MAKING

Using materials available in this classroom, make a demonstration that shows Newton's third law in operation. It will be more convincing if your demonstration uses objects that do not have the same mass. When you present, explain how your demonstration exhibits the third law.

Universal Law of Gravitation

ANALYZING

Suppose a planet as massive as the earth were shaped like a football. What effect would this have on people standing at various positions on its surface? Sketch this planet with various people standing on it (not to scale) indicate possible weight vectors, considering both magnitude and direction. Assume everyone has a mass of 100kg.

JUSTIFYING

Suppose you were on a journey between the earth and moon when you and your crew discovered two small, nearly identical asteroids floating 40 meters apart. Your crewmate Elsa analyzes the rocks and realizes that an astronaut could jump with enough force to get off one asteroid and fall onto the surface of the other, and then do the reverse to go back. She says an astronaut could keep this up for hours of fun. Jermaine disagrees. He says there would be problems with this. Who is right? Why? Be as detailed as possible.

MAKING

Create a new sport that can only be performed in high or low gravity. Give the goals, rules, and playing conditions of the game. Also indicate the gravitational field necessary to play this game, and specify the characteristics of the planet that would make this possible.

Perpendicular Force on a Moving Object, Direction and Speed

ANALYZING

Compare and contrast the motion of a satellite in circular orbit with a satellite in eccentric orbit. Make sure to mention the places in each where there is no change in speed. Why do these places exist, and what happens in the other places? Then, explain why the satellite is always accelerating in each orbit despite what happens to the speed.

JUSTIFYING

Hans and Bjorn are arguing about the shape of the course that would make their toboggan go the fastest. Hans says that a series of tight turns would make them go fastest. Bjorn insists that one, long smooth turn running the length of the whole track would be better—as long as the total length of both tracks is equal. Who's right and why?

MAKING

Using the plastic tracks in the lab, make a corkscrew course where the toy cars are always accelerating. Include places where the speed is increasing and where the speed is staying the same. Explain how you made each, including the places where there is acceleration, but where the speed is constant. Take measurements and make calculations that confirm you have achieved your goal. If we run out of track, sketch how you think the course would look, and explain your design.

Circular Motion and Centripetal Force

ANALYZING

Think of at least three situations where a constant force directed toward the center results in some object moving in a circle. What are the forces and what are the objects? What is similar and what is different in each situation?

JUSTIFYING

Melania says that it is impossible to swing a weight on a string in a circle without moving your hand a little bit. She says because of that, the earth must move in a small circle to make the moon orbit around it. Do you agree or disagree with her? Why? (Note to teacher: the earth does move in a circle, but because of system center of mass. This multiple-phenomena situation is my justification for making this a 5 instead of a 3 or 4. The debate might also lead to a discussion about the reason for making a circle with the string, and the physics associated with that.)

MAKING

Using materials in the classroom, make a set-up that either shows or disproves that a constant force is needed to keep an object moving at a constant speed in a circle.

Centripetal Acceleration

ANALYZING

Consider freeway off ramps and racetrack turns. These roads are designed for safety, making the road angled to minimize the role of friction in making the turn. Using your understanding of vectors and centripetal acceleration, find a formula for how much a road should be angled.

JUSTIFYING

Jerry says that there must be a reason the inner planets have nearly circular orbits while the outer planets tend to be more eccentric. Zelda says it's random. Who do you side with and why?

MAKING

Use what you know about centripetal acceleration to design a spaceship that simulates earth's gravitation. The period of rotation must be at least one minute. Make a sketch and diagram it with the most important information. Include bonus features that would open up more opportunities than those found on earth.

Relativity

ANALYZING

Derive the time dilation equation using the sequences of sketches depicting an astronaut moving near light speeds and shining a flashlight into a mirror. In addition, annotate each side of the resulting right triangle indicating what the side represents and the meaning of each variable that comprises the side. Use the terms "proper time" and "relative time" in your derivation. You may reference the formula sheet to see that your final form matches the sheet's.

JUSTIFYING

It is possible to "steal" energy from orbiting planets by having a spaceship creep up on a planet from behind and follow it in orbit for some time. The force of gravity between the planet and ship increases the speed of the ship in the direction of travel of the planet. Once the ship has sufficiently gained speed, it will orbit the planet at just the right distance to get flung into the desired direction, usually outward, away from the sun. Tanner argues that this method will help us get to relativistic speeds so we can travel to neighboring stars. Leticia says it can't. Who's right and why?

MAKING

Make a poem that rhymes and explains either the concept of time dilation or space contraction, or both.

Projectile Motion in Two Dimensions

ANALYZING

An Olympic ski jumper hits a horizontal ramp at high speed and is launched into the air until he touches down on the far side of a steep hill. Sketch a graph of his path in the air, putting vertical distance y on the vertical axis and horizontal distance x on the horizontal. You don't need to calculate numbers, but think of the approximate shape. Assume there is no air resistance.

On the same graph, sketch the shape of the path with air resistance. Then explain the reasons for the different shapes of the curves at various positions along the paths.

JUSTIFYING

Claudine and Flo are cliff diving. Claudine says she doesn't like hitting the water so hard, so she is going to run as fast as she can horizontally, which should take some speed off her vertical descent. Flo says that's silly. Who's right? Why? (Note to teacher: the increased speed will invoke more air resistance, which will slow her down, but. . .)

MAKING

Make a spreadsheet that will graph y vs. x for any projectile horizontally launched at any speed for any input gravity. For example, g on Mars = 3.73 m/s^2 and g on earth is 9.81 m/s^2. Assume the projectile is free to fall as far as 2000 meters, maybe more.

Vector Resolution

ANALYZING

Think of vectors that point in any direction in the x-y plane: force, velocity, momentum, etc. Compare how you resolve these vectors with how you get the values for the sine and cosine functions at various angles around the unit circle. Include comments about the four quadrants of the circle.

JUSTIFYING

Jenny says where you place a bomb in a building matters, because this affects the forces vectors of the explosion. Chuck says a bomb is a bomb. As long as it's in the building it will do damage. Who's right?

MAKING

Make a chart that allows you to quickly calculate vector component values by a simple application of unit circle values for the sine and cosine functions.

Balanced Forces, Statics

ANALYZING

Suppose you need to hang a weight between two buildings. You have a loosely hung rope connected to hooks directly across from each other. In the middle of the rope you tie a weight hanging from another rope. In order to make the tension in each of the two ropes the same, what angle would you try to achieve between the vertical rope and the rope attached to the buildings?

JUSTIFYING

Sheila works at a zip line service in a Panamanian jungle. She thinks the best ride would have the zip line be as tight as possible, so that the sagging line could be as straight as possible from one point to the next. Britney thinks that's a little dangerous, and not worth the effort. Who's right? Why?

MAKING

Use spring scales, string, and a mass to demonstrate the effect of hanging angle on string tension. Make a script and sequence of actions for your demonstration.

Inverse Square Laws, Gravity and Static Electricity

ANALYZING

Examine the geometry of light intensity vs. distance to explain static electric force vs. distance.

JUSTIFYING

Simone says that we can never really explore the universe because earth always pulls us back, no matter how far out we go. Support or refute his argument.

MAKING

Come up with at least two different real-life scenarios that roughly approximate the static electric and gravitational inverse-square relationships between magnitude and distance. Your scenarios may include interactions between people or interactions in nature. Explain why your two relationships are this way.

Forces and Motion, General Questions

ANALYZING

- How are these four speeds different?

  V_0 V_{const} V_{avg} V_{inst}

- If you were to do a lab to find Jerk, how would you find a place or make a situation of smoothly changing acceleration?
- Why do you think that a rectangular solid's small face has the same amount of friction as its large face when the object is dragged across a surface?
- Use Newton's 3 laws of motion to explain how a bomb explodes and does damage.
- Use Newton's laws of motion to explain how a rocket moves.
- Last week we smashed furniture with a sledge hammer. How did that demonstrate the first law of motion?
- How would you have to design buildings if friction did not exist?
- How are special and general relativity different?

JUSTIFYING

- Cables attached to a motor move elevators up and down inside a building. If the cable breaks, people usually die. However, some people say that if you could jump just before the elevator hits the bottom, you might survive. Do you agree with this idea? Why or why not?

- Will it be better to try a graphical method or pure calculations to find Jerk on the hill?

MAKING

- Design a blast shield that will protect you from an exploding bomb. Make a sketch and label its parts. Include the physics aspects that guide your design.

- Suppose you invent a material having a coefficient of static or kinetic friction that can be instantly changed by computer controlled remote signaling (Bluetooth or radio). The range can be $0.01 < \mu < 3.0$. Amazing! Now design a device or system that takes advantage of that feature. Sketches are welcome.

Chapter 6: Conservation Laws of Energy and Momentum

Kinetic Energy, $E=\frac{1}{2}mv^2$

ANALYZING

They say energy is the ability to do work. Show how kinetic energy and work can be related.

JUSTIFYING

Elba really loves baseball. She respects exceptional pitchers more than any other player. She says that the harder a pitcher throws, the bigger the ball's kinetic energy. This is important, because kinetic energy increases with the square of the speed. So a little more force goes a long way. Is she right?

MAKING

One of the theories about the formation of the moon is that a small-planet-sized object hit the earth and bumped a huge chunk into space. The chunk of matter was sprayed outward as smaller particles that orbited the earth and eventually came together to form one large body, the moon. Using measurements of the earth-moon system, estimate the size and speed of a planetoid that could have had the kinetic energy to cause such a result. Compare this energy to the kinetic energy of the earth as it orbits the sun. (Your answers can be based on rough estimates. For example, the gravitational field changes as you get further away from earth, so you can use constant or adjusted values for g.)

Potential Energy, $E=mgh$

ANALYZING

They say energy is the ability to do work. Show how potential energy and work can be related.

JUSTIFYING

Do you ever wonder how they drive poles into the ocean floor to build piers that we can walk on? It takes a lot of work to get them deep enough that they survive the turbulence of the tide and waves. The machine that does this is the pile driver. It uses a motor to slowly lift a heavy weight that is then dropped on the pole. The force drives the pole a certain distance into the ocean floor each time it is dropped. Cathy says that the pile driver can work equally well if it drops from a certain distance, or drops from twice that distance with half the number of drops. Rebecca thinks that's not the best way. Who's right?

MAKING

Hydroelectric power derives mostly from the potential energy of water that falls through a distance to turn turbines that generate electric power. Research the electrical needs of a small town, and the operation of a damn with hydroelectric power generation. From that, determine the annual flow rate in m^3 per second of a river that would be capable of powering that town.

Conservation of Energy

ANALYZING

Three things should be clear about energy: 1) for a falling object in the absence of wind resistance, the sum of the kinetic energy and potential energy should be constant, 2) Potential energy = mgh, while kinetic energy = $1/2mv^2$, and 3) the potential energy decreases proportionally to the distance dropped, easily seen in the formula PE = mgh, mentioned above.

However, what needs clarification is the speed and kinetic energy of a falling object. Therefore, analyze a situation in which a mass is falling. Consider speed vs. distance fallen, speed vs. time, kinetic energy vs. distance fallen, kinetic energy vs. time. Make a graph for each and comment on the relationships.

JUSTIFYING

The Yeti-kids is a club that does a lot of fun things. Each year they have a contest where they make toy cars out of blocks of wood and race them down a ramp. They are allowed to vary the design. One of the things they can do is add a certain amount of lead (a heavy metal) anywhere in their car. In their instructions booklet the Yeti-kids' leadership says it doesn't matter where in the car you put the lead; it won't affect the speed.

You are going to enter the next competition. Do you listen to the Yeti-kids, or do you see an advantage to ignoring them? Why?

MAKING

Using what you know about energy and centripetal acceleration, sketch a track for a roller coaster that will release a car from a high position, letting gravity speed it up and take it through loops that make the car go upside down. Of course, a roller coaster is fun when it goes fast and when there is centripetal acceleration involved. So, design the track for fun. Annotate at least 3 sections that are a speed you want and 3 sections that achieve a centripetal acceleration you want. Show the calculations for these values.

Momentum, p=mv

ANALYZING

Compare how falling objects of differing mass gain speed vs. how falling objects of differing mass gain momentum.

JUSTIFYING

For thousands of years people have asked a version of this question: "What happens when an unstoppable force meets an immovable object?" This is a paradox: We really can't have both. Considering the universe and what's inside it, do you think there is more likely to be an "unstoppable force," or an "immovable object," or something in between? What is your evidence for your belief?

MAKING

Make and test a hypothesis about the momentum of objects that orbit other objects at various distances. Your data could be existing information about mass, orbiting distance, orbiting period, and perhaps speed.

Conservation of Momentum

ANALYZING

For falling objects, what increases faster, momentum or energy?

JUSTIFYING

Caroline just toured the *Gimme Space* company's rocket making facility. She was impressed by the size of the rockets, almost 100m long! When she got back to school, she explained to Sara that the rockets had to be big and massive because the momentum they built up was needed to keep them going long enough to get out of earth's gravitational pull. Sara says Caroline misunderstands the relationship between momentum and energy. It is best for rockets to be as light as possible. Who's right?

MAKING

Using the 4 dm^2 sails, frictionless carts, masses and tracks, can you create a system that will maximize the wind energy acquired as your sailcar goes from one end of the track to the other? Will the design be different if you limit sail time to 3.0s? If you seek to maximize momentum, will your sail car system have a different design a) going the full length of the track, and b) with a 3.0s limit?

Impulse and Momentum Change

ANALYZING

If you are in a car accident, is it better to be in a big car or a small one? Why?

JUSTIFYING

In football, linemen are the biggest, strongest guys on the field because their job is to stop movement. On plays when the quarterback passes the ball, offensive linemen have to stop people from getting past them. These offensive linemen form sort of a wall. The defensive linemen try to get around that wall by pushing through or dodging their way around individual offensive linemen.

Both kinds of linemen have a lot of muscle and a lot of fat on them to make them strong and massive. Rex says defensive linemen tend to have the most muscle and not quite as much fat. Roy says it's the offensive linemen that are stronger. Who's right and why?

MAKING

Take one rubber band, one frictionless cart, and three masses. Connect the band to the carts with the hook in the table. Put together a launching system that will impart the biggest momentum to your combination of car and masses. (You choose the right number of masses.) Give evidence that your combination does the job.

Conservation of Momentum and Energy, Elastic and Inelastic Collisions

ANALYZING

One light and one heavy frictionless cart are connected by a stretched rubber band. The light car is on the left. The carts have Velcro patches between them so that they will stick together when they meet. After they collide, will they a) move to the left, b) move to the right, or c) stay where they met? Why?

JUSTIFYING

Aurelio clearly shows that with a specific mass combination a moving cart and a stationary cart can collide elastically and end up moving with the same speed, but in opposite directions. He says this is evidence that elastic collisions are better than inelastic collisions at conserving energy. Sal says that's not enough information to prove his point. Who's right?

MAKING

Take a set of frictionless carts with magnets and arrange a few collisions with different mass distributions to show that both momentum and energy can be conserved in completely elastic collisions.

Conservation of Energy in Systems, Capacitors and Springs

ANALYZING

If your physics teacher gave you a box of batteries, wires, capacitors, resistors, and multi-meters, how could you set up a circuit to demonstrate that conservation of energy holds for electricity? Describe what you would look for in one place on the circuit and what you would look for in another place on the circuit.

JUSTIFYING

Jolene wants to make her science project be about conservation of energy. She thinks it would be very interesting to show how food calories result in the same quantity of energy appearing in the motions of the person eating the food. Dawn says no, this would be too difficult, and would not give good results. Who do you think is right and why?

MAKING

Using rubber bands and other objects available around the lab, make a demonstration showing that mechanical energy is conserved.

Conservation Laws of Energy and Momentum, General Questions

ANALYZING

In terms of energy, explain why bicycles use up to ten or more gears, depending on the terrain you are riding on.

JUSTIFYING

- One form of muscular development exercises a few people used to do in the 1960's and 1970's is called "plyometrics". It is regaining popularity. Plyometrics uses impact to create spikes of force that are bigger than the weights a person can lift naturally, but the forces last for a very short period of time, much less than a second. For example a plyometrics leg exercise has athletes jump off a box. The person experiences a really big force in his thighs when he lands on his feet. Do you think plyometrics is a good idea? Why or why not?

- If you were in a car accident, would you rather it be a totally elastic or totally inelastic collision? Why?

- One style of martial arts says that every punch or strike should obey the rule: "hard on soft, soft on hard," which means that if you are going to hit someone on a bony feature like a chin, one should use the palm of the hand. If one is to attack the belly, he should use something hard, like knuckles or elbows. Is this a good rule? Justify your answer.

- Think of work, power, and momentum. For each, identify one sport in which winning performance is more a function of one of these concepts than the other two. Then justify your designations. For example, you might say, "(a particular sport) depends more on maximizing power than momentum or work, because. . . "

MAKING

As you know, *power* is *energy* divided by *time*. It would be interesting to know how much energy you or any human can muster in a given amount of time, say, 10 seconds. In one extreme, if you tried to lift an very heavy weight and got only one lift completed in 10 seconds, that would probably not be too impressive, in terms of power. In the other extreme, 30 lifts of a piece of paper would also not turn heads. How can you make a procedure to find your maximum power?

(male) traman
candy for trees bubble
prod (this protects anything that could hurt him from the sky)
poison bombs (these grow from the males prod)
giant root

Chapter 7: Thermodynamics

3.Energy cannot be created or destroyed, although in many processes energy is transferred to the environment as heat. As a basis for understanding this concept:

Energy Transfer Via Work and Heat Flow

ANALYZING

Think back on what happens during collisions. Momentum is always conserved, but energy is not. (Well, it is in a way, but not as measured in the final speeds of the colliding objects.) Elastic collisions do conserve energy, but inelastic ones do not.

A car crash is partially elastic and partially inelastic. Explain what happens to the “lost” energy when two cars collide. List all the ways energy is lost, and explain what ultimately happens to that energy.

JUSTIFYING

Sophie says a toaster is 100% efficient, because its purpose is to make heat. It does not need to do any work. Is she right? Why or why not?

MAKING

Make a system that does work when heat is added to it. Use a balloon filled with air as one part of your system.

Heat Engines and the First Law of Thermodynamics

ANALYZING

Compare an oil burning power plant with a hydroelectric plant. Where along each of the two processes can heat be wasted? How significant do each of these losses seem? Given the number and severity of these losses, which plant seems like it should be more efficient? Check your answers with actual, reported plant efficiencies.

JUSTIFYING

Rank these renewable energy sources by most feasible to least, giving justification for your answer: geothermal, wind, and tidal.

MAKING

Design a simple power plant based on some energy source we haven't yet considered. Sketch the plant showing the energy source, how it is captured, how it is transformed, and how it is delivered. Also annotate the diagram with the relevant symbols for the first law of thermodynamics, including arrows indicating the direction of energy flow into and out of the system.

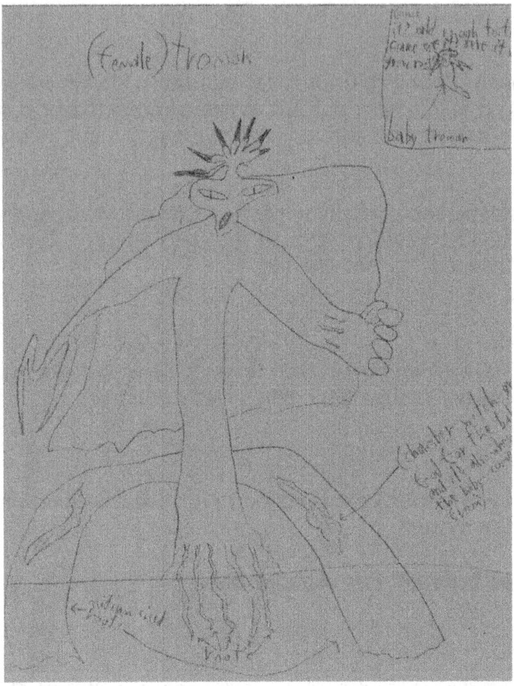

Thermal Energy and Sub-microscopic Motion

ANALYZING

Usually, the volumetric coefficient of thermal expansion is exactly three times the linear coefficient of thermal expansion. Why is this? Also, why is it that for some substances, the ratio is considerably less than 3?

JUSTIFYING

Clyde says that the phenomenon of thermal expansion proves that phase changes do not truly occur at a particular temperature. He says that because of the increased separation in molecules or atoms, potential energy changes are already beginning—even as temperature and kinetic energy rise toward the commonly accepted phase change temperature. Bonnie says Clyde has gone off the rails, and should just stick to the textbook on this issue. Who's right?

MAKING

Using materials available around the lab, make a thermometer. Over what range of temperatures do you expect your thermometer to work? How accurately does your thermometer respond to temperature changes? How does it work?

Natural and Forced Convection

ANALYZING

Identify one situation in which natural convection heats an object or a space and another situation in which natural convection cools an object or a space. For each, sketch an annotated diagram of the process. What are the similarities and differences between the two processes?

JUSTIFYING

It was a very hot day and the school had no air conditioning. Mrs. Sanders told her second grade class that they would cool off better if they stayed as still as possible and just let the air find them. Was she right?

MAKING

If you could redesign this classroom to take full advantage of natural convection, what changes would you make to improve heating, cooling, or both? Make an annotated sketch of the room with changes.

Tendency Toward Disorder

ANALYZING

Think about the ways we make our houses able to conserve as much heat as possible on a cold winter day. In writing, compare these methods with the statement, "most processes decrease a system's order, and eventually energy is spread evenly." Annotated sketches are welcome.

JUSTIFYING

Mr. Jones and Ms. Chen sit down for coffee at a physics teachers' conference. Mr. Jones is ready to give up on assigning labs that demonstrate the conservation of energy, because they never work right. The number of joules in the first situation is always more than the number of joules after the energy transformation. Ms. Chen disagrees. She thinks there are plenty of teaching opportunities in such labs. Who do you think is right? Why?

MAKING

Using two-liter bottles, shoe boxes or jars, along with materials from the lab or home, make two systems. One system should show a *decrease* in order after you shake it, and the other should show an *increase* in order after you shake it. Be prepared to compare your systems with the rest of the class' work. Observe which kind of system is generally more difficult to make, and how this relates to the second law of thermodynamics.

Entropy as a Measure of System Disorder

ANALYZING

Imagine two systems: in the first, a burst of steam into the system introduces a certain amount of energy. In the second, another burst of steam brings with it the same amount of energy. The first system is comprised of a gas at low temperature. The second system is comprised of the same gas, but at a higher temperature. Though both systems receive the same amount of input energy, which system experiences the bigger change in entropy? Give a logical explanation with sketches of what is happening at the sub-microscopic level in each system.

JUSTIFYING

Bertha and Bella are having a conversation about entropy in systems. Bertha thinks that if you get equal numbers of molecules of a particular gas, put them into two identical inflexible containers, and heat them to different temperatures, they will still have the same entropy because the spacing of the molecules is fixed by the container size. Bella says temperature still matters. Who is right and why?

MAKING

Take 9 pennies and place them on a tray in an ordered arrangement. Develop a method of measuring the degree of disorder, should an outside force affect the arrangement. Assume your starting quantity of disorder is zero. Then, drop a small rock into the tray from a height of 40 cm. Measure your degree of disorder. Then drop the same rock a second time from 40 cm height. Take a final measure of disorder. Explain, in terms of thermodynamics, what is happening.

As an extension, drop the rock four more times, measuring the disorder of your system each time. Graph the results of all seven states vs. energy added. Your energy added should be the accumulation of each equal-mgh quantity stemming from the rock drops.

Entropy and the Second Law of Thermodynamics

ANALYZING

Suppose you have a brother Bobby who has a strange medical condition. Every once in a while he has an attack. He forgets everything, even who he is, and just walks around randomly. He can open doors in this state. It is summertime, and under normal conditions, Bobby spends equal time in his room, in the house, and outside.

Regardless of where Bobby is when he gets an attack, where is he most likely to end up, his room, the house, or outside? Why? Refer to the second law of thermodynamics in your explanation.

JUSTIFYING

Ramón found out about the eventual heat death of the universe and became fascinated with entropy and how it increases. When he heard about tidal locking, he thought this might be an exception.

In tidal locking, a planet gets pulled into an oval shape by tidal forces from the body it is orbiting—or that orbits it. As a result, a planet orbiting a star and experiencing this reshaping might eventually stop rotating.

Ramón thought that if this happened to the earth, this would actually decrease entropy, since one side of the planet would have all the heat, and the other side would always face the darkness of space—the separation of heat would have more order. Plus, no motion would also imply more order.

Enrico laughed at Ramón, saying he wasn't thinking deeply enough and that a tidally-locked earth is a great example of the tendency of entropy to increase.

Who do you side with and why?

MAKING

Make a demonstration to teach middle schoolers the concept of entropy and its tendency to increase.

Heat Engines and Efficiency

ANALYZING

List ways we make cars more efficient in how we service, maintain and (possibly) modify them. How do these adjustments save energy? Explain for each item on your list.

JUSTIFYING

Kelly says that we have to stop making anything but the most efficient power plants. After all, if we know what kinds of energy sources have the least waste, using them would be both optimally economical and maximally earth-friendly. Rachel says we can sacrifice some efficiency for cleaner fuels. Who's right and why?

MAKING

Using materials in the lab, propose a modification to the hotplate that will maximize efficiency in its operation. Make annotated sketches of your changes, with explanations of the logic behind them. If there is time, make and test your design.

Thermodynamics, General Questions

ANALYZING

- How does the energy we learned about in the last unit connect to thermal energy?
- A 100 kg iron meteorite travelling at 25000 m/s slams into an airless moon. Why would such an impact not leave a clean hole? Give me general and specific thoughts.
- How might climate change affect the kinds of energy generation we do, not based on attempts to protect our weather, but considering the new opportunities changed weather might bring?
- Why does natural convection tend to not be sufficient for cooling animals and machines of increasing size?

JUSTIFYING

- Suppose the Gimme Space Company wants to make a space-suit-free region of Mars. They intend on doing this by using nuclear explosives to change the direction of an asteroid so that it will graze Mars' surface and leave a deep trench. The result would be that the surrounding (mostly carbon-dioxide) gas would fill-in to create a high-pressure region in the bottom of the trench, making it not necessary to be in a pressurized suit. Is this a good plan? Explain.

- Heather lives in North Dakota. She complains that no matter how many layers of clothes she puts on, she's always cold! Is entropy her problem? Explain.

- A few years ago chicken farmers came under pressure when citizens groups complained about the farmers feeding tens of thousands of LIVE chickens into wood chippers. Explain the difference in entropy that would have been created in the universe had the farmers first killed the chickens before feeding them into wood chippers.

MAKING

- Use a candle or a light bulb to demonstrate heat transfer by conduction, radiation, and convection. Make sure you have evidence that heat transfer is taking place. Which method do you think will be most effective? Follow safety procedures with the flame or current.

- How can you use a hair dryer to demonstrate the first law of thermodynamics?

Chapter 8: Waves

Waves as Energy Carriers

ANALYZING

List at least three kinds of waves, explaining the form of energy they carry and how they carry it.

JUSTIFYING

Coming out of a time of contemplation and insight, Elsa realizes that everything is a vibration. Therefore, she says, a wave is *the only way* to transport energy. Is she right? Why or why not?

MAKING

Suppose world governments in the future banned the use of electricity. Devise a system of communicating faster than the speed of sound in air. Make a sketch, annotate it, and explain how it works. Include an estimate for how fast the communication travels.

Transverse and Longitudinal Waves on Strings, Springs and the Earth

ANALYZING

What determines whether a medium can support a transverse, a longitudinal, or both kinds of waves?

JUSTIFYING

Xochi says she can generate both transverse and longitudinal waves in her jump rope at the same time. Do you agree this is possible? Why or why not?

MAKING

Make a demonstration that shows and explains both transverse and longitudinal waves.

Problems in Wave Speed, Wavelength, and Frequency

ANALYZING

Sketch a *displacement vs. distance* graph of a transverse wave using values you make up. Now sketch a *displacement vs. time* graph of the same wave, making up values as needed. What is all the information about the wave that you can get from the first graph? What information could you get if you just had the second graph? What could you know about the wave from both graphs?

JUSTIFYING

In the search for extra-terrestrial life, people have wondered why we don't pick up radio signals from other stars. It seems obvious that intelligent life would have learned to communicate by radio. Some say aliens *are* using radio, but have encrypted the signals so that other civilizations cannot identify them and come plunder their planets.

Sadiq thinks alien life is even more sophisticated, using extremely long wavelengths to communicate—much longer than radio, at a place in the electromagnetic spectrum that other beings would not likely pay attention to. Maya thinks that is an interesting concept, but thinks it would be more effective to use the other extreme end of the EM spectrum. Tisha thinks neither strategy would work. Who do you side with and why?

MAKING

Make your own system of classifying waves other than “transverse and longitudinal”.

Sound as a Medium-Dependent Longitudinal Wave

ANALYZING

Guitar strings get higher (in frequency) when you tighten them. Why is this?

JUSTIFYING

Georgina wants to protect her chameleon from too much noise, so she nests her lizard’s terrarium inside several larger glass enclosures, injecting different gasses inside each gap. She says these various transitions will kill the sound. Do you agree or not? Justify your answer.

MAKING

Earthquakes generate both longitudinal and transverse waves. Since the longitudinal waves move faster than the destructive transverse waves, house alarms can warn you to get out before the transverse wave hits. Research local faults and the speeds of the two kinds of waves in your area. Then, make a chart approximating the time spans between alarm and destruction where you live. Your chart should report on a number of earthquake points of origin.

The Electromagnetic Spectrum and the Speed of Light

ANALYZING

People often think about how it might be possible to make information travel faster than the speed of light. One idea is based on the idea of rotational velocity. For example, if you hold a meter stick in your hand and rotate it, the end of the stick will move twice as fast as the middle of the stick. If you had a long enough stick, the end could move really fast. If instead of a stick, you used a flashlight, and shined it way off into the stars, your rotation might make the end of the beam of light go faster than the speed of light. Show why this is not really possible.

JUSTIFYING

Stella says the equations of motion do not really apply to light. Do you agree? Justify your answer.

MAKING

Suppose colonies of humans were established around the few stars closest to earth. Make a map of these stars, including communication distances and times between them.

Properties of Waves: Interference, Beats, Polarization, Doppler Effect, Diffraction, and Refraction

ANALYZING

When you swim underwater, why does everything look blurry? And why does it look clear again when you wear a diving mask? Make an annotated sketch for each condition, with and without a diving mask. Use these sketches to help you explain.

JUSTIFYING

A large meteorite breaks up as it enters the atmosphere. It separates into three parts, each one landing in the ocean at the same instant in time. Apelu says the impacts will create waves that will interfere just like a two-slit wave experiment. Malosi says no, the three rocks will make a pattern significantly different. Who's right, Apelu or Malosi, and why?

MAKING

Design a situation in which the Doppler effect results in a smooth change in sound frequency.

Waves, General Questions

ANALYZING

- Find an example of when minimizing the critical angle is a good thing. Conversely, find an example of when maximizing the critical angle is a good thing. Compare the two systems.

- How do surfers "ride waves"? Are they being displaced? Are they a part of the wave? Are they part of the medium?

- How is it possible that guitar strings make noise when waves go from one end of the guitar to the other?

- Compare and contrast how lenses and fiber optics take advantage of the fact that combinations of two media have different refractive indices.

JUSTIFYING

- Tesla attached a motor to a steel beam in a building. The motor was imbalanced, with its rotational frequency tuned to match the resonant frequency of the building. As a result, the building eventually started shaking violently. Had Tesla not taken the motor off the building, would it have fallen down?

- Eric says his glasses give him headaches. He is looking into making his entire car windshield into a lens that matches his prescription. Is this a good idea?

MAKING

- Make a system of oscillators that demonstrate beat frequency. Be able to identify the beat frequency quantitatively. Don't use a demonstration you've seen before.

- How might you make corrective lenses that can be instantly adjusted to match a user's prescription?

- Arrange these glass and plastic pieces to make the fattest spectral dispersion band using this white light source and 0.3m square for your assembly.

Chapter 9: Electricity and Magnetism

Voltage and Current in DC Circuits Comprised of Wires, Batteries, Resistors and Capacitors

ANALYZING

Compare simple parallel and series resistor circuits to the water flow analogy by sketching a water circuit for each. Label the diagrams and explain how the analogy works. Remember, a water pump corresponds to a battery and a narrowed pipe represents a resistor. How might you represent a capacitor?

JUSTIFYING

Julie says you can take the plastic insulation off wires and make circuits that work exactly the same. The insulation is only a safety issue. Kelly says Julie is wrong. Having no insulation can change the circuit dramatically. Who's right and why?

MAKING

Make a circuit that lights a bulb when the switch is closed, allows the bulb to go dark, then lights the bulb when the switch is open again.

Ohm's Law

ANALYZING

When we draw circuits, we usually assume that wires have no resistance. We also say that resistors have a specific resistance. This makes it easy to apply Ohm's law for circuits. If we know the battery voltage, we can get the current. What's wrong with this picture?

JUSTIFYING

What is a better strategy for getting the circuit current flow you want, wiring batteries in series, or arranging a combination of series and parallel resistors?

MAKING

Are there social situations where people behave like they are following Ohm's Law? Of course there are! Give an example, and explain how it fits the relationships among voltage, current, and resistance.

Power in DC Circuits

ANALYZING

What do you think are important considerations when engineers design a toaster? Think about effectiveness and safety.

JUSTIFYING

Nikola Tesla had a vision for transporting energy through the upper atmosphere at very low resistance. This would be a great benefit to humankind, for obvious reasons. Is his idea a good one, or flawed? Justify your answer.

MAKING

Suppose you want a circuit expending 950W of power. Design a circuit to come as close as possible to this goal. Your materials are limited to wires, a 12 V battery and an arrangement of any or all of these resistors: 3, 4 ohm resistors; 2, 5 ohm, and 1, 1 ohm.

Electric Charges and Fields

ANALYZING

Compare patterns in electric field lines when there are at least two like charges with field line patterns when there are mixed charges.

JUSTIFYING

Irina has a problem with drawing electric field lines. She says in "real life" charges rarely stay still or last long enough for it to be worth trying to represent their electric fields in sketches. Do you agree or disagree? Why?

MAKING

Make a spreadsheet that calculates electric fields and/or forces when charge and/or distance is changed for two or more charges.

Sources of Magnetic Fields and Interacting Magnetic Fields

ANALYZING

You know how moving charges react to magnetic fields. How would one moving charge react to another moving charge? Imagine two scenarios where charges fly past each other (at different angles). Describe each charge's influence on the subsequent motion of the other.

JUSTIFYING

For her physics project, Ella wants to make a sensor that can detect the presence of live wires in the walls of her house. She says she could design her device to detect either the electric field or the magnetic field; it doesn't matter. Is she right?

MAKING

On paper, sketch several regions of magnetic fields so that if a charge enters from one side, its path traces an interesting pattern or shape. Be ready to explain how you did it.

Direction of Magnetic Field Produced by a Current

ANALYZING

When you arrange two wires sides by side and run DC current through each, you can cause the two wires to attract or repel each other depending on the directions of the currents relative to one another. What happens if you use AC current? What happens if the two AC currents are completely out of phase?

JUSTIFYING

Is it possible to stack coils of wires so that there is no magnetic field in a particular region around the coils even though current flows in the wires?

MAKING

On paper, design a system of current-carrying wires that results in an enclosed region of zero or greatly reduced magnetic field.

Changing Magnetic Fields Inducing Currents in Wires

ANALYZING

Recall the Lenz's Law demonstration in which a magnet dropped into a copper pipe induced eddy currents in the pipe. In that demo the magnet and pipe were round. What would happen if you repeated the demonstration with the same magnet, but this time you used a square pipe? Compare and contrast the nature of the eddy currents that might appear in both pipes and how these would affect motion of the magnet through the pipes.

JUSTIFYING

Ray thinks a sneaky way to steal electric power would be to make some loops of wire and place them near power lines in remote locations around the county. The ends of the loops could then be used to charge large storage batteries. Aside from the obvious moral issues, could this work? Justify your yes or no answer.

MAKING

How could you make a device that works like the Lenz's Law demo, but efficiently converts the gravitational energy of the dropping magnet into electricity? Make an annotated sketch of the device, and write a paragraph to explain how the device works.

Plasma, Form and Function

ANALYZING

Why would adding more energy to gases result in a separation of electrons from atoms? Why wouldn't the atoms come apart some other way? Why would they not come apart this way in the first place, when the substance changed state from liquid to gas?

JUSTIFYING

Natia says that when a gas heats to plasma phase, it becomes conductive, behaving much like metals. Metals, she says, have a "sea of electrons" that move easily under the influence of an electric field. Samaria says the comparison is not a good one, since metals are solids and plasmas are gaseous. Who's right and why?

MAKING

You know that the the transition from gas to plasma is characterized by a substance's liberation of electrons and/or ions, and therefore, high conductivity. Find or make an analogy from nature or society. Write it down.

Electric and Magnetic Fields and Lines of Force

ANALYZING

How is an electric field like a rubber band? How does it differ?

JUSTIFYING

Heather says force fields like those you find in science fiction movies behave very much like electric fields. For example, she says, a positively charged particle encountering an electric field will experience similar results. Defend or critique Heather's belief.

MAKING

How many different ways can you arrange charges so that they create an electric field that one non-moving positively charged particle cannot escape from? Sketch a few of these on paper. If all your sketches are two-dimensional, would any of your configurations work in three-dimensions? Make a final sketch that would absolutely trap the charge in three-dimensional space.

Force on a Charge in a Field

ANALYZING

How does the gravitational force compare to the electric force? Compare the mutual gravitational and electric forces on two protons at various distances. Compare also the field values for gravity and electric force at your various distances.

JUSTIFYING

Katie says we are more used to gravitational fields that are constant and electric fields that are changing simply because those are the most common manifestations here on earth, but both g and E can be similarly constant or changing. Do you agree or not? Why?

MAKING

Your teacher lost the pith balls he uses to demonstrate electric field. Choose materials and build a couple of replacement spheres. Design so that your system has the largest separation angle possible upon being charged. Explain and diagram how you did it.

Electric Field from One Charge

ANALYZING

Make a list of other physical relationships that behave like the inverse square dependence of E on point charge distance. Compare to simple inverse, direct, and square relationships in physics.

JUSTIFYING

If you could shape a planet such that it had a uniform gravitational field even over significant distances away from the planet, would its shape be similar to that of an electric plate designed for constant field?

MAKING

Design a tool that will show you the relative strength of an electric field at any position around a charged object.

Electric Fields from Multiple Charges

ANALYZING

If you wanted to create a strangely shaped electric field, would it be better to use a collection of point charges, or a strangely shaped conductor, which you would then charge? What are the advantages and disadvantages of each?

JUSTIFYING

Gravity accurately predicts the effects of masses on each other. However, in the small scale, microgravity is not that useful. Is electricity the opposite? In other words, do we benefit from the manufacturing of small scale electronic components while consideration of large scale electric fields is rarely necessary and mostly useless? Give your argument in support of one or the other viewpoint.

MAKING

Play with the electric field simulation.

Magnetic Force on a Moving Charge

ANALYZING

Using a magnetic field, how could you make a charged particle move in a spiral path?

JUSTIFYING

Since the earth is surrounded by a magnetic field, don't all moving charges travel along curved paths?

MAKING

Suppose you worked at a lunar facility surrounded by mountains. Suppose also that someone could shoot positive charges at you from any mountain. How could you design a circuit that could deflect any intruding positive charge upward, away from the central building of your compound? Sketch the circuit and explain how it works.

Conservation of Energy Featuring Electric or Gravitational Potential Energy

ANALYZING

Andrew, a carpenter, uses an electric motor to raise beams of wood two stories up, where he is building a wall. If we wanted to confirm that energy is conserved in this process, what would we measure? What would we compare? What are the possible wasters of unaccounted-for energy?

JUSTIFYING

If melted snow could flow without friction down a 2 kilometer high mountain, how fast would it be going at the bottom? Is it possible to capture most of that energy using dams and hydroelectric turbines at various positions down the face of the mountain? Explain.

MAKING

Assuming you had any tool or piece of lab equipment you could need, how could you design a method of converting energy from one form to a second form and from the second form to a third? Choose from *electrical*, *mechanical*, *thermal*, and *light* energy. Also, how could you measure the energy present in each form? Which energy conversion do you think would be most efficient? . . .least efficient? Why? Diagram your method of conversion alongside your method of measurement. On your sketch, answer the above efficiency questions.

Electricity and Magnetism, General Questions

ANALYZING

- Make sketches of the following wires and magnetic fields around them. Assume the same current is going through each wire. Annotate to indicate the differences in the fields, which should also be obvious from your sketches: 1) a straight wire, 2) a wire in a tight loop, 3) a wire in a loose loop, and 4) a wire in a tight coil of four loops.

- How does a transformer work?

- How do electricity and magnetism interact?

- Think of lightning and think of getting shocked by touching a doorknob. How are these two events the same and how are they different?

JUSTIFYING

Shelby says we could save a lot of money on electricity if we just captured the electricity of a thunderstorm. Her plan is to put giant capacitors in underground insulated chambers and connect them to lightning rods that are charged slightly opposite that of the clouds, to coax the lightning to strike. As soon as the lightning hits, it would charge one giant plate of her capacitor. A mechanism would react quickly to disconnect the lightning rod so that the charge would be captured. Can this work? Why or why not?

MAKING

Aside from examples already discussed in class, name three practical (or fantastic) uses of a circuit comprised of a charged capacitor and a wire loop.

Chapter 10: Nuclear Physics

The Nucleus

ANALYZING

Look up the diameter of an alpha particle, a gold nucleus, and a gold atom. Assuming Rutherford shot alpha particles at a gold film that was only one atom thick, what proportion of the alpha particles would be on a path to collide with (actually touch) the nucleus? If the film were four atoms thick, how does that change your answer? Why are there actually more deflections than the hit proportion you found?

JUSTIFYING

Mike says he can use the masses and magnetic frictionless carts to simulate a direct hit of an alpha particle on a gold nucleus. Can he? Justify your answer.

MAKING

Make a model that communicates the relative sizes of an atom's (average electron cloud) diameter vs. its nucleus' diameter.

Atomic spectra and emitted photon wavelength $E=hf$, $\lambda= hc/E$

ANALYZING

Compare the relationships between energy and frequency for light and for mechanical waves.

JUSTIFYING

Keeva says it is no accident that green is in the middle of the visible spectrum. She claims our eyes evolved to be able to detect the most important color to our survival: green. Most plants are green because they do not absorb that color. Chlorophyll cannot use it in producing energy for the plant, so it is reflected from the plant. Therefore, our range of functional vision centered on this important color because it guaranteed recognition. Do you agree with her logic? Why or why not?

MAKING

Assuming humankind begins colonizing the galaxy, and assuming we can create robots who will help us in our daily tasks, and assuming their eyes have a design capacity of only twice the human range of wavelengths, and assuming they can only sample from three set ranges of wavelength, which three wavelength ranges would you choose for robots and why?

Photoelectric Effect, $hf = \phi + E$

ANALYZING

In the equation for the Photoelectric Effect, $hf = \phi + E$, E can sometimes be represented as $1/2mv^2$ and sometimes as eV_o. What does each mean, and what would the conditions have to be if both $1/2mv^2$ and eV_o had a place in the equation?

JUSTIFYING

We get plenty of high energy photons from the sun. Could we just let sunlight shine on some low work-function metal plates, capture the current of electrons coming off them, and have a cheap source of energy? Why or why not?

MAKING

In two sketches, arrange the components of the photoelectric effect (metal plate, wires, etc.) so that in one case, the current could flow forever, as long as light continues to shine on the plate; and in the other case, current eventually stops, even if light keeps shining.

Half-Life and Decay Curves

ANALYZING

Graph the results of the following two functions for isotope mass *M*. Let time *t* run from 0 to 80. Compare the two functions and their graphs. Which function gives a nuclear decay curve using half-lives? How do you know?

$M = 800(\frac{1}{2})^{t/10}$

$M = 800((t+10)/10)^{-1/2}$

JUSTIFYING

You know a nuclear decay curve can be represented in half-lives. Can there be such a thing as a third-life or a tenth-life? Support your answer with facts.

MAKING

"Make" an isotope with either a very long or very short half life. Begin by reading the following method. Then thoughtfully choose your starting conditions to get what you want.

- Name some pretend isotope.
- Select a starting mass for graphing, like 2000g.
- Choose some time period represented by each roll of a die, for example, 3.0s, 100 min, or 5700 yr.
- Select some decimal fraction multiplier between 0.70 and 0.95
- Make up a logical condition like, "multiply the current mass by the multiplier only if the numbers 1, 2, 3, or 4 come up on the die."
- Make a T-chart of Mass(g) vs. Time(s,min,yr), roll the die, and fill the table.
- Graph what you have.
- Use the graph to find your isotope's half-life.
- Make a paragraph explaining why this process successfully results in a half-life (Burtson, "From Theory to Classroom to Fun").

Radioactive Decay and Decay Constant

ANALYZING

Look at several isotopes of some radioactive element (higher atomic number than Bi) and see of there is some relationship between isotope mass and half-life.

JUSTIFYING

Dieter thinks that exponential math is overkill. There are other ways to express radioactive decay over time besides exponential functions. Do you agree or disagree? Give evidence for your answer.

MAKING

Imagine having lighting rigged to a circuit that caused the light intensity to decay exponentially. Name three scenarios where you might use it. Describe the situations and the intended effects on the people involved.

Unified Mass Unit, Mass Defect and Nuclear Binding Energy

ANALYZING

Without doing calculations, compare a situation in which giant rocks separated in space came together to form a planet, to a situation in which neutrons, protons, and electrons came together to form an atom (if it could happen that way). What are the similarities and differences in the forces and energies, etc. involved?

JUSTIFYING

Some radioactive elements can have 10 or more isotopes. These isotopes can have radically different half-lives. Do you think there is a relationship between half-life and binding energy for the set of isotopes of a particular element? Justify your answer.

MAKING

Not every heavenly body is perfectly round. Clumps of rock or ice in space can have strange shapes until they get to be from 400 to 600 km in diameter. Aside from size, density and rigidity of materials make a difference, but eventually, as size increases, rocks become little planets. Are there similar transitions for atoms--not sphericity, but something else? Describe these and why they happen.

Fission and Fusion

ANALYZING

Compare natural radioactive decay with nuclear fission. How are they similar and different? Consider a Venn diagram or any method that best communicates your thoughts.

JUSTIFYING

A large defense contractor announced that they are close to developing a functional, efficient nuclear fusion reactor. Uncle Bob says he's going to load up on company stock. Should he? Justify your answer.

MAKING

Make a graphic that elucidates the fact that as stars begin fusing heavier and heavier elements, they are getting closer to dying. The binding energy chart sort of does this, but not clearly enough.

Nuclear Physics, General Questions

ANALYZING

In what ways is it important to life on earth that there is a certain high degree of radioactivity in the core of the planet?

JUSTIFYING

- Kirk the curmudgeon says it's foolish to think we could colonize Mars. It has no atmosphere to speak of and no magnetic field. Colonists would be sitting ducks, exposed to sub-atomic radiation from the sun and galaxy. Eager Karen is ready to sign up to be in the first wave of inhabitants. Who's got the right idea and why?

- Why does the wave model of light fall short?

MAKING

- Knowing what you do about x-rays, how could you improve the design of the safety blanket that they put over you at the dentist's office? Should it provide more protection, or can it be made simpler, or better in some other way? Provide annotated sketches of the existing and new designs. Include a paragraph that summarizes the changes.

- What do you think causes half-lives to be different?

Chapter 11: Physics, General Questions

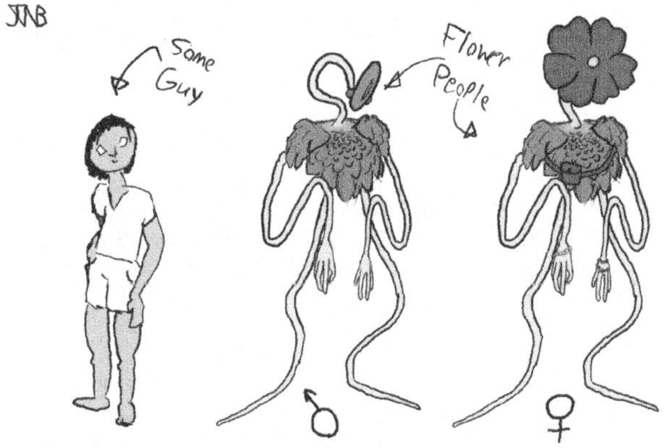

ANALYZING

- Make a Venn Diagram using these three categories: Vectors, Scalars, and Fundamental Units
- Make a Venn Diagram with these three categories: Motion, Waves, and Thermodynamics
- Suppose far in the future the mantle cooled enough that the earth was solid very far down. Suppose also that we discovered a method for opening up a huge cavern far below the surface of the earth. What would be some challenges to keeping that cavern stable and usable for gravitational experiments?
- Illustrate the design strengths of the teepee.
- Compare and contrast electricity and light.
- Some hobbies could benefit from a good understanding of physics. Make two lists, one of hobbies that benefit from practitioners knowing physics, and another of hobbies that don't. Include hobbies that are not obviously connected to physics. Be ready to defend your lists.
- Make a Venn diagram showing the overlapping and non-overlapping areas of physics and chemistry.
- How can a good understanding of physics help a person learn chemistry?
- How can a good understanding of physics help a person learn biology?

- How can a good understanding of physics help a person be better with home economics?

- Describe several ways trapeze artists depend on physics in their act (whether they are aware of it or not).

JUSTIFYING

- You take a time-machine trip far into the future, when the humans living on earth are dealing with a 30-hour long day, as the moon's gravitational pull continues to move us towards the deadly condition of tidal-locking. Humankind has finally mustered the will to deal with this, for future generations' sake. Four options gain popularity:

  a) Divert asteroids to impact the moon nudging it to a less influential orbit,
  b) Divert asteroids to impact the moon hard enough to break it into smaller pieces, which themselves can be similarly impacted to spread the moon into a kind of mini-asteroid belt equally distributed in orbit about the earth,
  c) Divert a large, moon-sized asteroid to a position around the earth's orbit so that it can cancel the existing moon's tidal bulging effects. (They may have to wait a few million years until one of these comes around.) Or,
  d) Do nothing.

  Each option comes with its own dangers and opportunities. Consider these. Then, after comparing the possibilities, indicate your choice and write to justify your decision.

- If you could live under a pressurized glass dome, what would be a more fun vacation place, the moon or Mars? Why?

- Where do you think most of our energy comes from now, and where will it come from in the future?

- Newton said,

  > That gravity should be innate, inherent, and essential to matter, so that one body can act on another at a distance, through a vacuum, without the mediation of anything else, by and through which their action and force may be conveyed from one to another, is to me so great an absurdity that I believe no man who has in philosophical matters a competent faculty of thinking, can ever fall into it (Niven, "The Scientific Papers of James Clerk Maxwell, vol 2," 316).

  Do you agree with Newton? Justify your answer.

- What is more exciting to you, theoretical or applied physics? Why?

MAKING

- Elon Musk proposed a system of transporting 20-30 people in a pod inside a 12-foot diameter tube. The pod would move at a speed of 700 miles and hour. How would you design the pod and the tube to meet these specifications and be safe and comfortable?
- Design a system that will trap a helium nucleus in a circle of radius 100m. Keep the speed of the nucleus less than 5% of the speed of light.
- What sorts of experiments would you perform in the sub-g low gravity conditions proposed in the deep cavern question above? Describe at least two experiments, including the *problem* or *question* and the *procedure* for each.

The end

Bibliography

Anderson, Lorin W., Krathwohl, David R., Airasian, Peter W., Cruikshank, Kathleen A., Mayer, Richard E., Pintrich, Paul R., Raths, James, Wittrock, Merlin C. *A Taxonomy for Learning, Teaching, and Assessing" A Revision of Bloom's Taxonomy of Educational Objectives, Complete Edition*. Boston: Allyn & Bacon, 2000.

Bloom, Benjamin, Engelhart, Max D., Furst, Edward J., Hill, Walker H., Krathwohl, David R. *Taxonomy of Educational Objectives: The Classification of Educational Goals*. New York: Longman, 1956.

Burtson, Eric. *From Theory to Classroom to Fun*. Publisher: Author, 2015.

Feynman, Richard P., Leighton, Ralph. *"Surely You're Joking, Mr. Feynman!": Adventures of a Curious Character*. New York: W.W. Norton & Company, 1997.

Harman P. M., *The Scientific Letters and Papers of James Clerk Maxwell Volume 1*.Cambridge: Cambridge University Press, 2009.

Llewellen, Douglas. *Inquire Within: Implementing Inquiry-Based Science Standards*. Thousand Oaks, CA: Corwin Press, Inc., 2002.

National Commission on Excellence in Education. *A Nation at Risk: The Imperative for Educational Reform*. Washington D.C.: U.S. Department of Education, 1983.

Niven, W. D., *The Scientific Papers of James Clerk Maxwell Volume 2*. Cambridge: Cambridge University Press; Reissue edition, 2011

Stigler, J.W., Gonzales, P.A., Kawanka, T., Knoll, S., Serrano, A. *The TIMSS Videotape Classroom Study: Methods and Findings from an Exploratory Research Project on Eighth-Grade Mathematics Instruction in Germany, Japan, and the United States*. Washington D.C.: U.S. Department of Education Institute of Education Sciences National Center for Education Statistics, 1995.

Stigler, James W., Hiebert, James. *Teaching Gap: Best Ideas from the World's Teachers for Improving Education in the Classroom*. New York: Free Press, 1999

Made in the USA
Monee, IL
04 November 2026

67387908R00046